ACADEMIC PLANNER
2022-2023

YEARLY GOALS 2022-2023

AUGUST 2022	FEBRUARY 2023
SEPTEMBER 2022	MARCH 2023
OCTOBER 2022	APRIL 2023
NOVEMBER 2022	MAY 2023
DECEMBER 2022	JUNE 2023
JANUARY 2023	JULY 2023

IMPORTANT DATES

WEEK TIMETABLE

TIME	S	M	T	W	T	F	S

ASSIGNMENT LOG

ASSIGNMENTS	CLASS	DUE DATE	GRADE

ASSIGNMENT LOG

ASSIGNMENTS	CLASS	DUE DATE	GRADE

ASSIGNMENT LOG

ASSIGNMENTS	CLASS	DUE DATE	GRADE

MY NOTES

MY NOTES

AUGUST 2022

IMPORTANT THIS MONTH

GOALS

DATES TO REMEMBER

MY NOTES

AUGUST 2022

SUNDAY	MONDAY	TUESDAY	WEDNESDAY
	1	2	3
7	8	9	10
14	15	16	17
21	22	23	24
28	29	30	31

NOTES:

AUGUST 2022

THURSDAY	FRIDAY	SATURDAY	TO DO
4	5	6	______________
11	12	13	______________
18	19	20	______________
25	26	27	______________
			______________

NOTES:
____

____

SEPTEMBER 2022

IMPORTANT THIS MONTH

GOALS

DATES TO REMEMBER

MY NOTES

SEPTEMBER 2022

SUNDAY	MONDAY	TUESDAY	WEDNESDAY
4	5	6	7
11	12	13	14
18	19	20	21
25	26	27	28

NOTES:

SEPTEMBER 2022

THURSDAY	FRIDAY	SATURDAY	TO DO
1	2	3	
8	9	10	
15	16	17	
22	23	24	
29	30		

NOTES:

OCTOBER 2022

IMPORTANT THIS MONTH

GOALS

DATES TO REMEMBER

MY NOTES

OCTOBER 2022

SUNDAY	MONDAY	TUESDAY	WEDNESDAY
2	3	4	5
9	10	11	12
16	17	18	19
23	24	25	26
30	31		

OCTOBER 2022

THURSDAY	FRIDAY	SATURDAY	TO DO
		1	______________
6	7	8	______________
13	14	15	______________
20	21	22	______________
27	28	29	______________

NOTES:

NOVEMBER 2022

IMPORTANT THIS MONTH

GOALS

DATES TO REMEMBER

MY NOTES

NOVEMBER 2022

SUNDAY	MONDAY	TUESDAY	WEDNESDAY
		1	2
6	7	8	8
13	14	15	16
20	21	22	23
27	28	29	30

NOTES:

NOVEMBER 2022

THURSDAY	FRIDAY	SATURDAY	TO DO
3	4	5	______________
10	11	12	______________
17	18	19	______________
24	25	26	______________
			______________

NOTES:

____

____

DECEMBER 2022

IMPORTANT THIS MONTH

GOALS

DATES TO REMEMBER

MY NOTES

DECEMBER 2022

SUNDAY	MONDAY	TUESDAY	WEDNESDAY
4	5	6	7
11	12	13	14
18	19	20	21
25	26	27	28

NOTES:

____

____

DECEMBER 2022

THURSDAY	FRIDAY	SATURDAY	TO DO
1	2	3	______________
8	9	10	______________
15	16	17	______________
22	23	24	______________
29	30	31	______________

NOTES:

____

____

JANUARY 2023

IMPORTANT THIS MONTH

GOALS

DATES TO REMEMBER

MY NOTES

JANUARY 2023

SUNDAY	MONDAY	TUESDAY	WEDNESDAY
1	2	3	4
8	9	10	11
15	16	17	18
22	23	24	25
29	30	31	

NOTES:

JANUARY 2023

THURSDAY	FRIDAY	SATURDAY	TO DO
5	6	7	
12	13	14	
19	20	21	
26	27	28	

NOTES:

FEBRUARY 2023

IMPORTANT THIS MONTH

GOALS

DATES TO REMEMBER

MY NOTES

FEBRUARY 2023

SUNDAY	MONDAY	TUESDAY	WEDNESDAY
			1
5	6	7	8
12	13	14	15
19	20	21	22
26	27	28	

NOTES:

FEBRUARY 2023

THURSDAY	FRIDAY	SATURDAY	TO DO
2	3	4	
9	10	11	
16	17	18	
23	24	25	

NOTES:

MARCH 2023

IMPORTANT THIS MONTH

GOALS

DATES TO REMEMBER

MY NOTES

MARCH 2023

SUNDAY	MONDAY	TUESDAY	WEDNESDAY
			1
5	6	7	8
12	13	14	15
19	20	21	22
26	27	28	29

NOTES:

MARCH 2023

THURSDAY	FRIDAY	SATURDAY	TO DO
2	3	4	
9	10	11	
16	17	18	
23	24	25	
30	31		

NOTES:

APRIL 2023

IMPORTANT THIS MONTH

GOALS

DATES TO REMEMBER

MY NOTES

APRIL 2023

SUNDAY	MONDAY	TUESDAY	WEDNESDAY
2	3	4	5
9	10	11	12
16	17	18	19
23	24	25	26
30			

APRIL 2023

THURSDAY	FRIDAY	SATURDAY	TO DO
		1	
6	7	8	
13	14	15	
20	21	22	
27	28	29	

NOTES:

MAY 2023

IMPORTANT THIS MONTH

GOALS

DATES TO REMEMBER

MY NOTES

MAY 2023

SUNDAY	MONDAY	TUESDAY	WEDNESDAY
	1	2	3
7	8	9	10
14	15	16	17
21	22	23	24
28	29	30	31

NOTES:

MAY 2023

THURSDAY	FRIDAY	SATURDAY	TO DO
4	5	6	__________
11	12	13	__________
18	19	20	__________
25	26	27	__________
			__________

NOTES:
____

____

JUNE 2023

IMPORTANT THIS MONTH

GOALS

DATES TO REMEMBER

MY NOTES

JUNE 2023

SUNDAY	MONDAY	TUESDAY	WEDNESDAY
4	5	6	7
11	12	13	14
18	19	20	21
25	26	27	28

NOTES:

JUNE 2023

THURSDAY	FRIDAY	SATURDAY	TO DO
1	2	3	
8	9	10	
15	16	17	
22	23	24	
29	30		

NOTES:

JULY 2023

IMPORTANT THIS MONTH

GOALS

DATES TO REMEMBER

MY NOTES

JULY 2023

SUNDAY	MONDAY	TUESDAY	WEDNESDAY
2	3	4	5
9	10	11	12
16	17	18	19
23	24	25	26
30	31		

JULY 2023

THURSDAY	FRIDAY	SATURDAY	TO DO
		1	______
6	7	8	______
13	14	15	______
20	21	22	______
27	28	29	______

NOTES:

CONTACT PAGE

NAME	NUMBER

CONTACT PAGE

NAME	NUMBER
____________________	____________________
____________________	____________________
____________________	____________________
____________________	____________________
____________________	____________________
____________________	____________________
____________________	____________________
____________________	____________________
____________________	____________________
____________________	____________________

CONTACT PAGE

NAME

NUMBER

CONTACT PAGE

NAME	NUMBER

CONTACT PAGE

NAME	NUMBER

PASSWORD PAGE

WEBSITE	PASSWORD

PASSWORD PAGE

WEBSITE

PASSWORD

PASSWORD PAGE

WEBSITE	PASSWORD

PASSWORD PAGE

WEBSITE	PASSWORD
_________________	_________________
_________________	_________________
_________________	_________________
_________________	_________________
_________________	_________________
_________________	_________________
_________________	_________________
_________________	_________________
_________________	_________________

PASSWORD PAGE

WEBSITE	PASSWORD

MY NOTES

MY NOTES

MY NOTES

MY NOTES

MY NOTES

MY NOTES

MY NOTES

MY NOTES

MY NOTES

MY NOTES

MY NOTES

MY NOTES

MY NOTES